Before Starting a SharePoint Workflow Project

Permissions Handling

TIPS, Ideas & Solutions

By: Hilal Al Sukairi

ISBN-13: 978-1542557283

ISBN-10: 1542557283

TABLE OF CONTENTS

About The Author

Hilal Al Shukairi is a long time developer, who gained an experience by involving in many development projects for the last 14 years. He has worked as a SharePoint workflow developer for more than one year that he spent most of his time in studying and solving critical issues. This is the time to share his experience, findings, and solutions with you.

Introduction

- Is SharePoint a good choice for my project?
- Can I handle the SharePoint permissions automatically by the workflow?
- What should I do and what should not?
- What is the best practice to do it?

The answers vary from the person who learned about it or experienced it. The power of practicing is more. Unlike other SharePoint Books, this book is written based on an experience gained from an actual project plus education such that it provides the best-practice tips, ideas, and solutions. You may take hours to read this book, but it will save you days, weeks and even months in your SharePoint workflow project.

This book is a unique book such that it focuses on the major experience points which have never been studied in any available book. These books mostly touched every single point in the SharePoint either from technical or end-user perspective. But they missed some points which cannot be discovered unless by working in an actual project.

This book is designed not to be too much technical, but it is a collection of issues, ideas, and solutions which help you to accelerate your SharePoint workflow project or even to decide whether to go with SharePoint or not. Furthermore, this book version focuses mainly on the issues of handling permissions on SharePoint using the workflow.

I wrote this book because I had a moment in a big project that I said: "I hope that I knew these points and ideas before I start the project, which has been delayed by months even though I have all the required knowledge in this field."

Finally, this book not dedicated for a certain SharePoint version because it focuses on Ideas that can be utilized in any version. Even though, this experience is gained mainly from working with SharePoint 2013.

Intended Audience

This Book is for you if you are:

Intermediate Developer: Facing a lot of unexpected issues while developing the workflow. If you still not starting yet the project will be better such that you utilize this book Ideas and techniques.

Project manager: Not sure whether SharePoint is a good choice for your project or not.

Software architecture: Tries to design a SharePoint Permission structure for the project.

However, the technical information provided in this book is assuming you already know about the SharePoint and its terms. Also, it assumes that the developer already knows how to implement the SharePoint workflows and activities.

Why to Choose SharePoint

The answer is easier when the question is why not to choose SharePoint? Such that SharePoint from its name and marketed features attract to be choosing for many projects expecting that it will be very straight forward and easy to be utilized. In Fact, it is not always easy or works as expected. Therefore, this topic can give you a better understanding whether to go with SharePoint or not.

The major facts to be considered about SharePoint are:

- **It is not a SharePoint but better to be called a collaboration point**: it designed to help the project team members to collaborate and share documents in an efficient way which guarantees the last version of the document to be available and accessible.

- **It is not a Document Management System (DMS) but a File Management System**: The concept of (DMS) deferred than the concept of collaboration software like SharePoint. Basically, from Access control perspective, The DMS is designed for a very complex access controls such that it restrict access to some sensitive documents like legal documents and financial report which you cannot access except by having the right permission. On The other hand, SharePoint is designed to provide accesses to documents in a "project-based" where the project team members can access and share their documents utilizing the built-in version control. However, if you are trying to use SharePoint as a document management system, then this book helps you to a certain scope.

- **It is not fast enough:** SharePoint is not as fast as an ASP.net project since it does a lot more work like security checking and getting files and data from the database.

- **It is not the friendliest user interface**: needs training mostly.

- **It is not a database, but it stores data as lists and libraries** which better to considered as excel sheets that can connect to other sheets.

- **It is not an ASP.net but needs special skills**: The developers required to have a special training and skills to start developing SharePoint applications and workflows even though the SharePoint is based on ASP.net, the expert on ASP.net cannot start developing in the SharePoint environment without that special knowledge. Moreover, the development environment is slower and more difficult to debug than normal ASP.net project.

- **It is not cheap**: the SharePoint cost may not exceed 7% of the actual cost required to implement it. As beside the SharePoint license, there are SQL server licenses, Windows Server licenses, MS office 365 license if needed, IT resources, backup solution, Servers hardware, users training, anti-viruses … etc.

- **It is not unlimited but has boundaries:** the limitations and boundaries have to be considered before you start designing a SharePoint Project. An example of that is the number of security principals for each object limited to 2000 max. Therefore, may be a good design consideration to use group principals rather than user's principals for any securable object if the number of targeted users is huge.

For sure, SharePoint has a huge list of Features which comes out of the box, and it can be utilized for your projects like site provision, search, workflow, and alert. However, listing the previous major points which SharePoint is not, just to clarify and remove the confusing of mixing between the SharePoint features and other full systems. Clear Ideas and correct thoughts are the best way to help the decision maker to decide whether to utilize SharePoint or not.

SharePoint Workflow

SharePoint workflow helps on creating automatic or manual tasks to the SharePoint objects like lists, list items, pages, …etc. It automates the business processes and can be programmed for a specific task.

How can it be programmed?

There are two ways to program your custom Workflow for the SharePoint, either by using the SharePoint designer (including importing from MS Visio) or Visual Studio (SharePoint Workflow Project). This book is focusing on the visual studio workflow projects. Since visual studio has a lot of extra features and abilities than the SharePoint designer.

How the workflow works?

The workflow can be started either manual or based on an event. Therefore, it can be attached either to the whole site or a specific list or library. If the workflow attached to a list or library it can be started by the following:

- Manual start.

- On an item is inserted.

- On an item is updated.

What facts should you consider about the Workflow?

- **It works as an independent service from SharePoint**: monitoring the health of this service is important such that if it is not working for any reason, then the workflow will fail to start even though the SharePoint is working.

- **It is not immediate action; but it takes time to complete the task** (milliseconds, seconds or even minutes depends on the workflow complication and other components).

- **It requires special application permission**: It treated as an application where the administrator can give it special access such that to work with the SharePoint objects.

- **It works with the user permission:** beside the application permission, if the workflow started by a user, and the user does not have the authorization to access the targeted object, then the workflow not has that too.

- **You need to design your SharePoint activities**: The workflow project in Visual Studio is unlike the other kind of projects which you find that Microsoft supplies it with great controls such that you do not need to have your custom control for it. Like, for example, the data grid, text box, data model …etc. In SharePoint workflow projects, the activities considered as those controls which you can reuse them and drag them to your project. The supplied workflow activities in the Visual Studio are too much general. Which can be utilized to do anything but it is complicated on a big project. This book recommends you to prepare custom activities which make your tasks to be done easily and fast before starting your workflow project. Also, this book describes 15 abstract activities you should implement, which help you to do complicated permission handling tasks in the easiest way possible.

- **It can wait for other events after it started:** the workflow has wait activities which can be utilized to let the workflow pauses until an event completed. These events for example like waiting for a flag on a list to equal to "1".

- **The workflow is not started automatically with the SharePoint system account:** If the workflow designed to start on inserting or updating an item. Then if that item is inserted or modified by a system account, the workflow is not triggered.

Rest Vs CSOM

The SharePoint is designed to allow integration with other applications. The integration and the communication are possible using REST or CSOM APIs. In fact, Visual Studio supports the workflow project with special activities to work with the SharePoint object like getting the list, items and doing some task like inserting, updating and deleting items. However, as mentions before, it is very general and does not help to do some complicated tasks like handling permissions. However, the workflow can do a much more special task using REST or CSOM.

REST API is very easy and straight forward and has a great documentation from Microsoft. However, in some cases, there are undocumented parts that REST API cannot do as described. Then you should use CSOM which is a little bit more complicated. From an Experience point of view, use the REST API for most of your tasks. On the other hand, for the cases that REST API is not working or no REST API to do that task then use CSOM. However, you do not have to learn CSOM. I use a technique where I utilize (Fiddler web debugging software – See Extra Resources "2") which can capture any communication between any application and the SharePoint. I did that by using the SharePoint Designer where I copy the HTTP post sent in between. I just then modify the required parameters and use the prepared XML in my activity. This way save you much time and apply the same technique with any application you use other than the SharePoint Designer. Thus, if that application can do the task then for sure you can.

Since this book focuses on permission handling, Microsoft has a great summarizing PDF file for the REST API (check Extra Resources #3) which shows you all the functions in a graphical way. You only need to know how to utilize them and what cases they are not working as expected.

The following are some REST functions that not work as expected in some cases where you should use CSOM:

- **addRoleAssignment (principalid, roledefid):** this REST API Work great but only if the principal already exists with any role on that object (on RTM version of SharePoint 2013). Therefore, you cannot add for example a reading role to a user to read a list item if that user has not existed on that object. This principal is either a user or group. However, by using CSOM, you can add it to that object, and it is added automatically to the site root as Limited Access role.
- **CreateSharePointGroup**: you cannot create a SharePoint group with an owner assigned to it using REST API. However, you can do that with CSOM.
- **addPrincipalToGroup**: You cannot add any user to a group if you are not a group owner or a SharePoint admin. Also, since the workflow is running by the user who started it, then that user must be an owner of the group to be able for the workflow to control that group. Good news, you can assign a group to be the owner of any group in SharePoint. Therefore, I recommend creating a special group as the owner of other groups and put on it any user you want to be able to add or remove any user.

SharePoint Permission Structure

Permissions Level

There is a set of default permissions level in SharePoint (See Extra Resources #4). These permissions level are just predefined roles that can be assigned to a principal to authorize it to a certain level of access like (read, contribute). These default permissions can be edited by the admin to add or remove some permissions except the "Limited Access" and "Full Control" are uneditable. Moreover, "limited access" permission is assigned automatically by the SharePoint when you add any principal. For example, if you add a "read" role for a user to read a list item. Then it is added to the list automatically as "Limited Access".

One of the most important permission levels you need in your project is "Limited Access". This permission level gives the principal access to the object without any extra permission. This level is the least level of the permissions. However, it is important such that you do not want the principal to do anything extra to the list other than accessing it without being able to read any item on it but at the same time you want to give the user a "Contributor" permission in one of the list item. That is why SharePoint adds this level of permission automatically to the list when you try to give the user any permission level in one of its items.

On the other hand, what about if you want to give that user a "limited access" to a list as default. So the user can access the list but without being able to see any items on it except when you give it the right authorization to one of the list items. Unfortunately, the "Limited Access" permission cannot be programmatically assigned because it is done automatically by SharePoint. The solution is easy; you can create your custom "Limited Access" permission by only cloning the original one, and you can name it like "My Limited Access" which then you can assign it to any principal just as the other permission levels.

The Default Permissions

The project designer should be aware of the project and how it should be constructed. Sometimes the project is too much complicated for the SharePoint. However, by understanding all the business flow in the project and how the permission flows between the users, that make you able to construct default permissions for any parent object. All the objects, in fact, inherited its initial permission from its parent. If the list item created, its default permission inherited from the list permissions.

Moreover, some cases are too complicated in such way that you have a deferent set of users who utilize the same list, but you cannot give the same initial permission for all of them. In this case, you can utilize the "Folders". They act as a parent of any items that inserted inside them.

More important is to consider the kind of document you deal with in the project for you to decide how to structure your permissions. Are you dealing with very sensitive documents in one list, and you should not share them with anyone except with certain users? In this case, the default list permission is to have only "Limited Access" to all the users. That means that by default they can not read any document inserted to the list unless they granted access to it by a certain business process. On the other hand, if your main concern is to share all the documents with a group of users for a certain period, then after that they cannot see them. In this case, Your default permission level for the list should be "Read" level assigned to this group such that any document added to the list they able to read it. Whereas, another process run at the end of the period to remove that read permission.

Inherited Permission and Unique Permissions

Any object in SharePoint is either inherits its permission from the parent or has a unique permission. The default for any object created is inheriting from its parent. Moreover, you cannot assign any unique permission to that object till you "Break its Inheritance". You can do that using REST or CSOM APIs. However, if you would like in any case to return it to the inheritance state, then you can do that by "Resetting its Inheritance". Moreover, you can do that by REST or CSOM APIs as well. Taking into consideration that when you break the object inheritance, you have the flowing options:

- **Copy Role Assignment:** If it is true then you copy the permission from the parent. Otherwise, you just give the permission to the current user.
- **Clear Sub Scopes:** if true, then all the child permissions will be cleared. For example, breaking inheritance of a List and this parameter is true will force all the list items to inherit its permission from the parent list even if that item has a unique permission. But if false then no child secure object permission will be changed.

Workflow Main Required Activities

The custom workflow activities help you breaking apart the complex workflow task to be easy. It is worth and highly recommended that you spend a few days implementing the activities required for your workflow project. However, you need a lot of experience while you are diving in the project until you discover all the activities you need to implement. As promised, this book guides you to start handling permissions easily using the SharePoint workflow by providing 15 main activities in abstract level such that for you to implement before starting your project.

This book expects that you already knew how to create custom activities for the SharePoint workflow. However, this great tutorial from (Jason Lee's Blog – See Extra Resources #1) can help you to start with some of the listed custom activities.

The structure of the listed custom activities required helps you to implement it fast by providing the inputs, outputs, expected results, and the functionality. Moreover, it includes the main REST API required to implement it or how you can prepare the CSOM query.

On the other hand, the listed activities categorized into REST APIs activities, CSOM activities, and Bonus activities. This categorizing letting you navigate easily to the category you most interested on even though it is recommended to walk through all of them since they are all important.

REST API Activities

1. Name: GetRoleDefinitionIDByName

Inputs:

- **RoleDefName:** The name of the role to get its ID. Eg: READ, Contributor, ..etc.

Outputs:

- **RoleDefId:** The role ID.
- **responseStatus:** the HTTP response content to verify the status of the request if success or there is an issue.

Result:

It returns the Role ID of the required role.

Main Required REST API Functions:

---

http://serverName/[site]/_api/web/RoleDefinitions/getByName('**RoleDefName**')

Benefits:

Most of the REST functions available in SharePoint REST API or CSOM require the ID of the role as a parameter rather than the name. Therefore, you will need this activity to provide the required role ID easily. Even though the pre-defined permissions levels have a fixed ID, but you don't have to store or memorize them. Moreover, the custom roles you created you will not be able to get their IDs except by this activity.

2. Name: BreakInheritanceForObject

Inputs:

- **objectPath:** the path to the secure object.
- **isCopyRoleAssignment:** true or false parameter, to determine whether to keep the inherited permission if it true. Otherwise, it gives the permission only to the current user.
- **isClearSubScope:** true or false parameter, to clear all the child permissions or to keep them.

Outputs:

- **responseStatus:** the HTTP response content to verify the status of the request if success or there is an issue.

Result:

It will break the inheritance or assign a unique permission to the secure object.

Main Required REST API Functions:

---

http://serverName/[site]/_api/web/**objectPath**/breakRoleInheritance(**isCopyRoleAssignment,isClearSubScope**)

Benefits:

Breaking the object role inheritance will allow it to be assigned any role for any user. No object could be assigned any unique role if it is not broken inheritance.

3. Name: ResetRoleInheritanceForObject

Inputs:

- **objectPath:** the path to the secure object.

Outputs:

- **responseStatus:** the HTTP response content to verify the status of the request if success or there is an issue.

Result:

It will reset the inheritance for the secure object. That means the object will be assigned the same permissions assigned to its parent all the time. Hence, it will not affect any child objects which have a unique permission.

Main Required REST API Functions:

---

http://*serverName*/[site]/_api/web/**objectPath**/resetRoleInheritance

---

Benefits:

Resetting the inheritance will allow you for some cases to return that object to the default permission which its parent has.

4. Name: addRoleForObjectToPrincipal

Inputs:

- **RoleDefId:** the ID of the role you want to be assigned.
- **objectPath:** the path to the secure object.
- **PrincipalId:** the principal ID of the user or group.

Outputs:

- **responseStatus:** the HTTP response content to verify the status of the request if success or there is an issue.

Result:

It will assign the user or group the role required to access the secure object.

Main Required REST API Functions:

---

http://serverName/[site]/_api/web/**objectPath**/RoleAssignments/AddRoleAssignment(***PrincipalId,RoleDefId***)

Benefits:

This activity is very useful regarding not creating complex steps to assign any permission level to any object. It makes the life very easy in such way that you need only to provide the required parameters.

Note: the above rest API function is not working as documented in RTM release of SharePoint 2013. It required the principal to already has some form of permission on that object. Normally, I use CSOM to assign a limited access role to that principal, and then I can use this activity to assign it any further roles. However, it is working fine in SharePoint online.

5. Name: RemovePrincipalRoleFromObjectPath

Inputs:

- **RoleDefId:** the ID of the role.
- **objectPath:** the path to the secure object.
- **PrincipalId:** the principal ID of the user or group.

Outputs:

- **responseStatus:** the HTTP response content to verify the status of the request if success or there is an issue.

Result:

It will remove from the user or group the role required to access the secure object.

Main Required REST API Functions:

http://serverName/[site]/_api/web/**objectPath**/RoleAssignments/removeRoleAssignment(**PrincipalId,RoleDefId**)

Benefits:

Removing any role from a secure object is so easy by this activity. It requires only providing the principal ID, object path, and the role definition ID to be removed. Providing the role definition ID is important because the user could have more than one role assigned, but you need only to remove a specific role.

6. Name: GetGroupIdByGroupName

Inputs:

- **groupName:** The name of the group.

Outputs:

- **groupId:** The group ID.
- **responseStatus:** the HTTP response content to verify the status of the request if success or there is an issue.

Result:

Returns the group ID.

Main Required REST API Functions:

---

http://serverName/[site]/_api/web/siteGroups/getByName(***groupName***)

Benefits:

Most of the REST API functions request the principal ID of the group you want to apply the function to. Thus, this activity provides the required information easily by the group name.

7. Name: addUserToGroup

Inputs:

- **userName:** The user name.
- **groupId:** The group ID.

Outputs:

- **responseStatus:** the HTTP response content to verify the status of the request if success or there is an issue.

Result:

It will add the user to the group.

Main Required REST API Functions:

1. Use this REST URL:

---

http://serverName/[site]/_api/web/siteGroups/getByName('**groupName**')/Users

---

2. Request Content: a dynamic variable with the following Properties:

   a. **path:** _metadata/type , **value:** SP.User

   b. **path:** LoginName , **value: userName**

Benefits:

Adding a user to the group is normal process by utilizing this activity.

8. Name: removeUserFromGroup

Inputs:

- **userName:** The user name.
- **groupId:** The group ID.

Outputs:

- **responseStatus:** the HTTP response content to verify the status of the request if success or there is an issue.

Result:

It will remove the user from the group.

Main Required REST API Functions:

1. Use this REST URL:

---

http://serverName/[site]/_api/web/siteGroups/getByName ('**groupName**')/Users/removeById(**userId**)

---

2. Get the **userId** from the **userName**.
3. Request Content: a dynamic variable with the following Properties:

   a. **path:** _metadata/type , **value:** SP.User

   b. **path:** LoginName , **value:** **userName**

Benefits:

Removing a user from a group is an easy task by utilizing this activity.

9. Name: CreateSharePointGroup

Inputs:

- **groupName:** The name of the group.

Outputs:

- **responseStatus:** the HTTP response content to verify the status of the request if success or there is an issue.

Result:

It will create a new SharePoint group with the name provided.

Main Required REST API Functions:

1. Use this REST URL:

http://serverName/[site]/_api/web/siteGroups

2. Request Content: a dynamic variable with the following properties:

a. **path**: _metadata/type , **value**: SP.Group

b. **path**:Title , **value**: **groupName**

Benefits:

It creates a SharePoint group with the group name. You can utilize it easily. However, if you need to add an owner to the groups then use the CSOM activity.

10. Name: createFolderInLibrary

Inputs:

- **libraryName:** The name of the library.
- **folderName:** The name of the folder.

Outputs:

- **responseStatus:** the HTTP response content to verify the status of the request if success or there is an issue.

Result:

It will create a new folder in the library.

Main Required REST API Functions:

1. get the **serverRelativeUrl** for the library root folder by the following

    a. use this REST URL:

---

http://serverName/[site]/_api/web/List/getbyname(**libraryName**)/Rootfolder

---

    b. Get the following property from the response content: "d/ServerRelativeUrl".

2. Then use this REST URL:

---

http://serverName/[site]/_api/web/folders

---

3. Request Content: a dynamic variable with the following properties:
   a. **path**: _metadata/type , **value**: SP.Folder
   b. **path**: ServerRelativeUrl, **value**:
      *(***serverRelativeUrl** + "/" + **folderName***)*

Benefits:

Creating a folder inside a library helps to provide deferent default permission for the library items. For example, you can create a Department_1 folder and have default permission for (department 1) employees to access it. Then any items added to this folder will inherit the folder permission by default.

CSOM Activities:

CSOM has a special XML format to be prepared. As mentions before, we will utilize the web debugging software to get the right XML for this function purpose. I have used "fiddler" software (Extra Resources #2). Hence, the following parameters needed for any CSOM activity:

- **siteGUID**: this founded by following:

  1. Request the following URL:

---

http://serverName/[site] /_api/site

---

  2. Get the following property: "d/Id" from the response content. And assign it to the "**siteGUID**".

- **webGUID**: this founded by following:

  1. Request the following URL:

---

http://serverName/[site]/_api/web

---

  2. Get the following property: "d/Id" from the response content, and assign it to the "**webGUID**".

In addition, the only CSOM URL requested is:

---

http://serverName/[site]/_vti_bin/client.svc/ProcessQuery

---

11. Name: CreateSharePointGroupWithOwner_CSOM

Inputs:

- **groupName:** The name of the group.
- **ownerId:** The principal ID of the owner group.

Outputs:

- **responseStatus:** the HTTP response content to verify the status of the request if success or there is an issue.

Result:

It will create a new SharePoint group with the name provided and assign a group owner for it.

How to do it:

This is the fastest way to do it:

1. Open The SharePoint Designer.
2. Use "Fiddler Software" to get any data between SharePoint server and the Designer.
3. Add a new group with a group owner to the SharePoint using the Designer.
4. In fiddler, Copy the HTTP request sent from the designer to the server.
5. Find (:site:) in 2 places and replace them and the next value with:

(:site: " + **siteGUID** + @")

6. Find (:web:) and replace it and the value next to it with:

   (:web: " + **webGUID** + @")

7. Find (:g:) and replace it and the value next to it with the group owner ID:

   (:g: "+ **ownerId** + @")

8. Find ("Name=""Title"" Type=""String"">) and replace it and the value next to it with:

   ("Name=""Title"" Type=""String"">"+**groupName**+@")

9. Utilize this xml in your activity and done.

Benefits:

This activity allows you to create a group with an owner. The power of it is to allow you to assign a group owner to the group. Because of the fact, you cannot add any user to the group unless if you are the group owner or the SharePoint admin. Therefore, it is more efficient for you to add all the users you want to manage this group to the group that owns it.

12. Name: AddRoleToUserUsingCSOM

Inputs:

- **RoleDefId:** the ID of the role you want to be assigned.
- **userName:** the username.

Outputs:

- **responseStatus:** the HTTP response content to verify the status of the request if success or there is an issue.

Result:

It will assign the user the role required to access the SharePoint.

How to do it:

The fastest way to do it is:

1. Open The SharePoint Designer.
2. Use your capture software "Fiddler" to get any data between SharePoint server and the Designer.
3. Add a new user to the SharePoint using the Designer.
4. In Fiddler: copy the HTTP request sent from the designer to the server.

5. Find (:site:) in 2 places and replace them and the value next to them with:

   (:site: " + **siteGUID** + @")

6. Find (:web:) in 2 places and replace them and the value next to them with:

   (:web: " + **webGUID** + @")

7. Find (:rd:) and replace it and the value next to it with the Role definition ID:

   (:rd: " + **roleDefID** + @")

8. Find ("Name=""EnsureUser""><Parameter Type=""String"">) and replace it and the value next to it with:

   ("Name=""EnsureUser""><Parameter Type=""String"">"+**userName**+@")

9. Utilize it in your activity and done.

Benefits:

This activity needed when the user does not exist with any role in the target SharePoint object. Especially, when the REST API that used for this purpose is not working properly in some cases.

13. Name: AddRoleToGroupUsingCSOM

Inputs:

- **RoleDefId:** the ID of the role.
- **groupId:** the principal ID of the group.

Outputs:

- **responseStatus:** the HTTP response content to verify the status of the request if success or there is an issue.

Result:

It will assign the group the role required to access the SharePoint.

How to do it:

The fastest way to do it is like the previous activity but here with a group instead of user. You will notice that only small changes on the copied XML than the previous.

1. Open The SharePoint Designer.
2. Use your capture software "Fiddler" to get any data between SharePoint server and the Designer.
3. Add a new group to the SharePoint using the Designer.
4. Copy the HTTP request sent from the designer.

5. Find (:site:) in 3 places and replace them and the value next to them with:

   (:site: " + **siteGUID** + @")

6. Find (:web:) in 2 places and replace them and the value next to them with:

   (:web: " + **webGUID** + @")

7. Find (:rd:) and replace it and the value next to it with the role definition ID:

   (:rd: " + **roleDefID** + @")

8. Find (:g:) and replace it and the value next to it with:

   (:g: "+**groupID** + @")

9. Utilize it in your activity and done.

Benefits:

This activity utilized when the REST API used for this purpose is not working properly. It would allow you to assign a group the right role even if it did not exist before in that object.

Bonus Activities

14. Name: getCollectionFromRestURL

Inputs:

- **restURL:** The URL that returns a collection of values with deferent properties.
- **propertyToRetrieve:** a string that defined what property you want to retrieve.
- **isString:** Boolean that determines whether the returned property has integer or string value.

Outputs:

- **intOutputCollection:** the stored integer values of the requested property if **isString** is false.
- **stringOutputCollection:** the stored string values of the requested property if **isString** is true.

Result:

Return either an integer or string collection of a specific property from a REST URL.

Main Required REST API Functions:

1. REST URL: Any URL that returns a list of values.
2. Foreach loop to go through the items and insert the specified property value to the integer or string collection.

Benefits:

This activity simplifies the task of retrieving a collection of values from the XML response of the REST API. Many cases can utilize this activity such as getting the list of user IDs of the SharePoint or a specific object, getting the list of all the group's names available, getting the group usernames, etc. In sum, any cases where you need to get a list of any property returned by the REST API.

15. Name: collectionsComparitor

Inputs:

- **ListA:** The first list in the comparison.
- **ListB:** The second list in the comparison.

Outputs:

- **AnotBList:** the return collection that has all the values which are in A but not in B.
- **BnotAList:** the return collection that has all the values which are in B but not in A.

Result:

Return two collections which are (A not in B collection) and (B not in A collection).

How to do it:

1. No REST API here just comparison by doing the following:
2. The foreach loop that goes through all the items in list A which checked if they are in B and add which are not in AnotBList.
3. The foreach loop that goes through all the items in list B which checked if it is in list A. Then adds any item not in A to BnotAList.

Benefits:

My favorite activity that it easy to implement but has valuable benefits. The major use of it is in synchronization. You can synchronize between for example a list of user in a SharePoint list and a group users. To do this example let say you have the following:

- You want the department's group to have the same users in the department list (sync)
- You assign ListA: the user IDs of all the employees in DepartmentAList.
- You assign ListB: the user IDs of all the users in the group which called departmentA.
- Then the results from this activity will be **AnotBList** and **BnotAList**.
- The synchronization is simply done by (adding the entire user IDs in **AnotBList** to the group) and (Removing all user IDs in **BnotAList** from the group) and done.

In Sum, each one of these activities will save your time by reducing the complicity. If you complete all the 15 activities before starting your project then handling permissions in SharePoint will be a playing game. You just need to take the major time in professionally designing and structuring your permissions based on your project requirement.

External Resources

Since this book is not a step by step book which explains how to do the ideas and solutions. This list of recourses will help you to apply what is mention in this book if you are not aware of them.

1. **SharePoint custom workflow Activity tutorial:** A great tutorial helps you to start creating your custom SharePoint Activity in **Jason Lee's Blog.**

URL: **www.jrjlee.com/2014/01/custom-workflow-activity-for-granting.html**

2. **Fiddler (The free web debugging proxy for any browser, system or platform):** this software is a great software let you utilize CSOM API in SharePoint by capturing any other application requests.

URL: **www.telerik.com/fiddler**

3. **SharePoint REST Syntax PDF:** This PDF from Microsoft it summarizes a lot of SharePoint REST API functions:

URL: **www.microsoft.com/en-us/download/confirmation.aspx?id=41147**

4. **Default permissions level:**

URL: **technet.microsoft.com/en-us/library/cc721640.aspx**

www.ingramcontent.com/pod-product-compliance
Lightning Source LLC
Chambersburg PA
CBHW070135210526
45170CB00013B/1106
* 9 7 8 1 5 4 2 5 5 7 2 8 3 *